Food
115

快乐的机器人

The Happy Robot

Gunter Pauli

[比] 冈特·鲍利　著

[哥伦] 凯瑟琳娜·巴赫　绘

颜莹莹　译

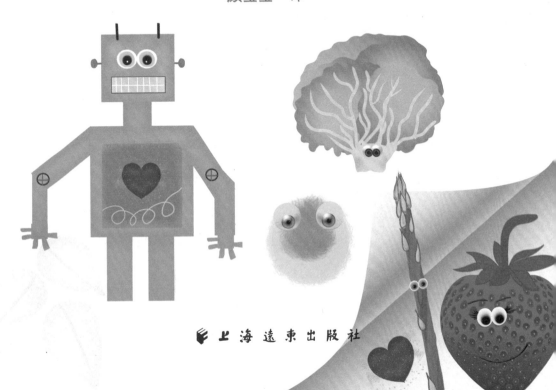

上海远东出版社

目录

Contents

一个黄色黏菌细胞正飘浮在空中。风把它越吹越远，这时它发现一棵小生菜正在快乐地生长，享受着充足的水和美食。

"这是个安家的好地方吗？"黏菌问。

"是的，这里很棒，我们的幼苗能很快从土壤中汲取营养，农户太会选地方了。"

A single yellow slime mould cell is floating through the air. As it is carried far away by the wind, it spots a small lettuce growing happily, enjoying enough water and good food.

"Is this a good place to settle down?" asks Mould.

"Yes, it is. This is a remarkable place where our seedlings can quickly access food from the soil. The farmer made a good choice."

······发现一棵小生菜······

... spots a small lettuce ...

... we can be turned into vegetable wire ...

"你的叶子很健康，很强壮。"

"我们不管是在地上还是地下都长得很快，人们认为可以把我们变成蔬菜电线。"生菜说。

"蔬菜电线？做什么用的？"

"You have such healthy, strong leaves."

"We are so good at growing, above and under the ground, that people think we can be turned into vegetable wire," Lettuce says.

"Vegetable wire? For what?"

"嗯，我们是天然的电线，富含水和纤维。我们可以传输养分，还可以传输电。"

"你能传输电？那是不是意味着人们可以不用再开采矿山制造铜线，只要种植你们就行？"

"完全正确！尽管人们现在还只是刚开始了解其中的原理，但是你能想象种植电线吗？"

"Well, our natural wire is full of water and fibres. We transport food, but we could also transport electricity."

"You could? That means people can stop mining all that ore and instead of having copper wire just grow you?"

"Exactly! People are only now starting to figure out how it works, but can you imagine, growing wires?"

……我们还可以传输电。

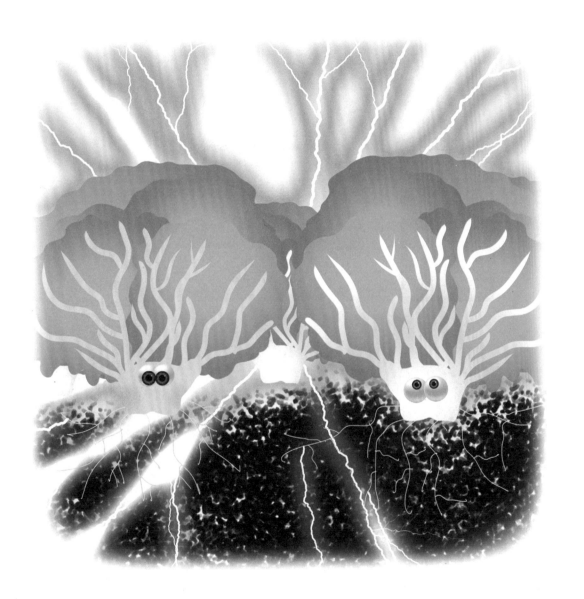

... we could also transport electricity.

……不需要电池，就可以为他们的玩具充电？

... power their toys, without any need for batteries?

"那什么时候才能准备好？"

"我们希望尽快。但只要人们还依赖那些巨大的高压电缆，我们能发挥的作用就会很微小。你知道吗？我们已经能够测量温度，并把信息传递给农户，让他知道我们的脚很冷。"

"这太不可思议了！你们可以用在玩具上吗？你能想象吗？孩子们不再需要电池，只要自己种植电线，就可以为他们的玩具充电。"

"When will this be ready?"

"Soon, we hope. But as long as people rely on those huge cables with very high voltage, our role will remain small. Do you know, we can already measure the temperature and pass that information on to the farmer to tell him that our feet are cold."

"That is incredible! And could you be used in toys? Imagine kids growing their own wires to power their toys, without any need for batteries?"

"这是个好主意！在我的种子发芽后三天，我就有了很好的连通能力，可以通过一个互相连通的传感器网络为人们提供我身边的一切信息。"

"我们真应该多聊聊，我们是同行，或许我做得更多些。"

"可如果你只是一个单细胞，要怎么才能变成电线传递信息呢？"

"What a good idea! Only three days after my seed has sprouted I am already well connected and ready to provide people with all the information around me, through a network of connected sensors."

"We should talk more as I am doing the same, and perhaps even a bit more."

"But how can you if you are just a single cell? How can you turn into a wire sending out information?"

……如果你只是一个单细胞？

... if you are just a single cell?

......一个巨大的个体。

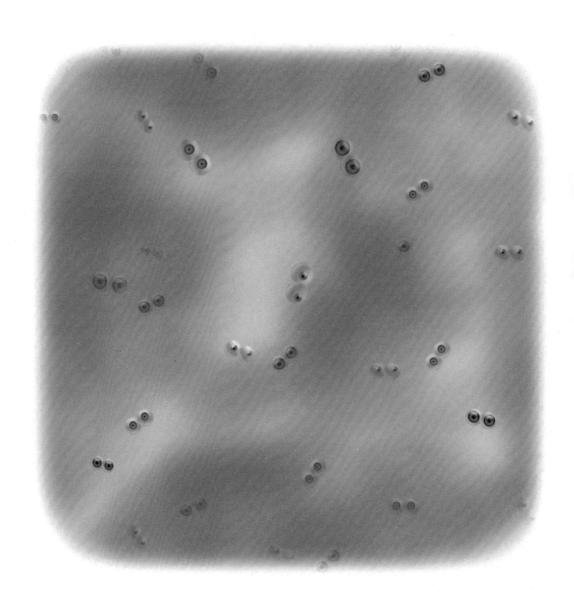

... a big, giant body.

"嗯，当食物充足的时候，我是喜欢独处的。可是当食物短缺、无法自给自足时，我们这些黏菌细胞就会聚集在一起，形成一个巨大的个体。"

"这听起来像恐怖电影里的东西。"

"拜托！我们最大也就长成几平方米大小。"

"那很大了，你都可以一口吞掉一整棵大生菜了。"

"Well, when there is a lot of food, I like to live alone and enjoy it. But when there is not enough for me to survive on my own, we mould cells get together and form a big, giant body."

"That sounds like something from a horror movie."

"Come on! We grow to, at best, only a few square metres in size."

"That is huge. You could gobble up a whole big lettuce all at once."

"我们这么做可不是为了征服世界！"黏菌说，"我们是为了找食物。"

"可是你甚至都没有大脑！你怎么会如此聪明呢？"

"瞧，人们知道，我们黏菌聚在一起是为了找食物，并且我们能记住食物在哪儿。"

"我是一棵植物，"生菜说，"没法到处走动。我必须为我所拥有的感到高兴，也为人类、鸟类和蜜蜂给予我的感到高兴。"

"We are not doing that to take over the world!" Mould says. "We do it to get food."

"But you don't even have a brain! How can you be so clever?"

"Look, people know that when we moulds work together we find food – and we remember where it is."

"I am a plant," says the lettuce, "so cannot move around. I have to be happy with what I do have, and what people, birds and bees give me."

我必须为我所拥有的感到高兴……

I have to be happy with what I do have ...

......可以告诉农户天有多冷......

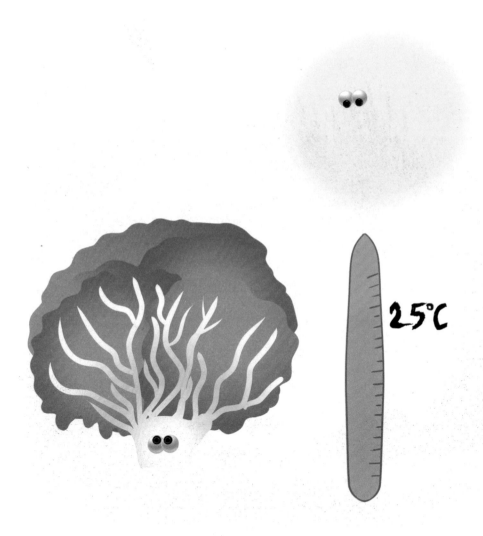

25℃

... can tell the farmer how cold it is ...

"嘿，你可以告诉农户天有多冷，而我可以唱歌，还能让机器人微笑。"

"什么？"生菜说，"不可能，你扯得有点远了吧。你能唱歌吗？像我一样唱歌？你又是怎么让机器人微笑的？"

"如果植物可以唱歌，为什么我不能呢？"黏菌反问道。

"好吧，但那样还是不能让机器人感到高兴！"

"Hey, you can tell the farmer how cold it is, and I can sing and make a robot smile."

"What?" Lettuce says. "No, now you are taking it a bit too far. Can you sing, like I can? And how do you make a robot smile?"

"If plants can sing, why can't I?" Mould wants to know.

"Okay, but that does not yet make a robot happy!"

"瞧，你的电线可以将我与机器人连接。我们可以让他因为太阳光太强而看上去很难过，或者让他因为食物很美味而微笑。"

"是的，我们可以一起让这个世界变得更快乐，甚至连机器人都微笑。"

……这仅仅是开始！……

"Look, your wires could connect me to the robot. We can make him look sad because the sun is too bright or make him smile because the food is good."

"Yes, together we can make this world a happier place, where even robots smile."

... AND IT HAS ONLY JUST BEGUN! ...

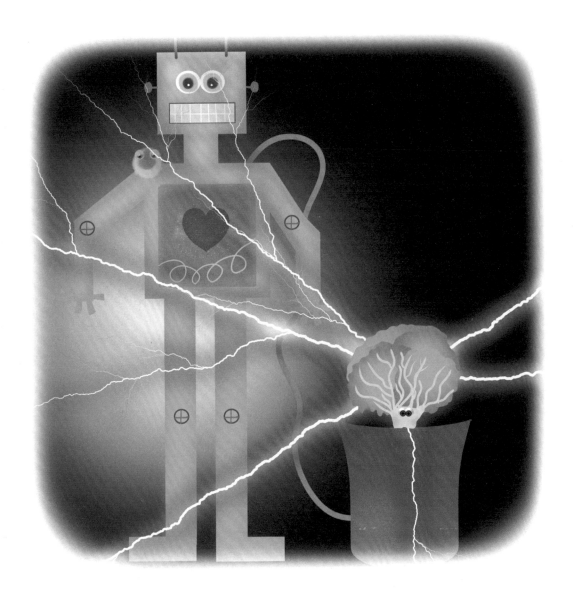

……这仅仅是开始！……

... AND IT HAS ONLY JUST BEGUN! ...

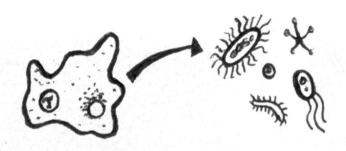

Slime moulds live alone when there is enough food. When there is a shortage of food, they will congregate and live as a single body. They feed on bacteria, yeast and fungi.

当食物充足时黏菌独自生存。当食物短缺时它们便会聚集成一个独立的细胞团。它们以细菌、酵母菌和真菌为食。

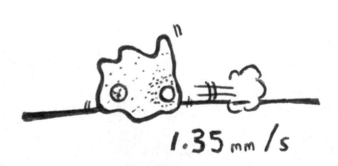

1.35 mm/s

The slime mould can move its body at a speed of 1.35 millimetres per second, about one centimetre an hour, which is the fastest of any micro-organism. Slime moulds reproduce when food is in short supply.

黏菌能以每秒 1.35 毫米的速度移动，1 小时移动约 1 厘米，是微生物界速度最快的。当食物短缺时黏菌就会繁殖。

Almost all living creatures, including humans, are able to conduct electricity and can therefore be used as "wires" to relay information, provided these "wires" remain motionless and do not degrade over time. Plants can also conduct electricity, provided they get enough light, water and minerals.

几乎所有的生物，包括人类都能够导电，可以像"电线"那样传输信息，前提是这些"电线"保持不动，且不会老化。植物也可以导电，只要它们有足够的光、水和矿物质。

Bio-scientists can create electronic components that could ultimately lead to bio-robots, based on our understanding of the behaviour of plants and slime moulds.

基于对植物和黏菌行为的了解，生物学家能够制造电子元件，进而制造出生物机器人。

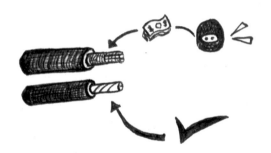

铜具有很好的导电性能。铝电缆的电阻是铜电缆的两倍。铜的高价值和易熔炼的特性使其成为偷盗的目标。铝价格低廉，不易熔炼，是不错的替代品。

Copper has very good electrical conductivity. An aluminium cable has twice the resistance of a copper cable. The high value of copper and ease of smelting made it a target for theft. Aluminium's low price and the great difficulty of smelting make it a competitive alternative.

只需要洒少量水，抓一把土即可种植植物，它们可以充当有机电线，为刚起步研发的新一代生物电路、传感器及信息处理器提供基础支持。

Plants can act as organic wires that could become the infrastructure behind a new generation of biological circuits, sensors and information processors that grow from scratch, needing little more than a sprinkle of water and a handful of soil.

植物并没有一个中央大脑，但是仍然能够感知、学习、记忆甚至像人一样有所反应。植物能感知地心引力和水的存在。它们还能在根部尚未接触到障碍物之前就感知到障碍物，进而改变根部的生长方向。

Plants do not have a centralised brain and yet can sense, learn, remember and even react in ways that would be familiar to humans. Plants can sense gravity and the presence of water. They can also sense when there is an obstacle in the way of their roots before making contact with it, and change the direction in which the roots grow.

植物没有神经细胞，但是有一套传递电信号的系统，能够产生与人类大脑类似的神经递质。

Plants do not have nerve cells but do have a system for sending electrical signals, and produce neurotransmitters similar to those in the human brain.

Think About It

想一想

Do you think that one day we will be able to grow electrical wires using plants, and have bio-robots?

你觉得将来有一天我们能够种植出植物电线，制造生物机器人吗？

Can plants think and feel? If so, do you consider them intelligent?

植物能思考和感知吗？如果能，你觉得它们聪明吗？

How long does it take to produce a metal cable, from the initial decision to open a mine until the cable is installed in a home? How long does it take for a plant to grow a local wire?

从一开始决定要开采一座矿山到最后电缆安装到户，制造一根金属电缆需要多长时间？种植一根植物电线需要多长时间？

Can robots be happy? Or do you think only people feel happiness?

机器人会感到快乐吗？还是你认为只有人类能感受到快乐？

Ask your teacher for an electric conductivity (EC) meter. Collect some fresh soil from a local vegetable garden and allow it to air-dry for a few hours. Now place it in a cup and add the same volume of distilled water. Stir it and let the suspension rest for half an hour. Stir it again and measure the EC. What is your conclusion: too much or not enough, or just the right amount of nutrients for the vegetables to grow?

向你的老师要一台电导仪。在当地的菜园里收集一些新鲜的泥土，自然风干几个小时。把泥土放在一个杯子里，加入等量的蒸馏水。搅动之后静置半小时，等悬浮物降落。再次搅动并测量电导率。你得出什么结论？太高还是不够？或者是养分刚好让植物生长？

学科知识
Academic Knowledge

生物学	生菜是一年生或二年生植物，菊科；生菜的茎干富含膳食纤维，其清淡可口的叶子富含维生素及矿物质；土壤水的电导率是监测农作物对养分的吸收程度的很好指标；电导率低说明植物需要更多的养分；当土壤变干，养分（主要是盐）含量升高，植物的根部会因盐的浓度太高而受到损害；植物的理想电导率在每厘米1到3毫西门子之间。
化 学	生菜富含维生素A和钾；合成化肥的主要原料是可溶性盐类，如氨、磷、钾、钙、镁或硫酸盐。
物 理	电导率是物质传导电荷能力的测量值，以西门子/米为测量单位，电流是单位时间内电子的流动；植物不能吸收过多的水分，基质中过多的水分会迫使空气溢出，导致根部因缺氧而腐烂。
工程学	水压与电压类似，而水流大小与电流大小类似，位于管道的两点间用来测水压的流量限制器与电阻器类似；测量电导率的电导仪用于监测在水耕栽培、水产养殖和淡水供应用水中的营养物、盐或杂质。
经济学	植物通过从土壤中摄取营养物质而生长，电导率在其中起到很大作用；使用化肥，可以使上述作用更加有效，这已经成为一个巨大的产业；使用更多的化肥并不能保证更多的产量。
伦理学	积极销售肥料往往导致过度使用，特别是不溶性品种，在水体中积累到一定程度就会引起水葫芦的过度繁殖以及赤潮的爆发。
历 史	1826年，乔治·西蒙·欧姆提出了著名的"欧姆定律"；1941年，艾萨克·阿西莫夫，一位以科幻和科普作品闻名于世的俄裔美籍作家，在小说《我，机器人》中首次使用了"机器人心理学"这一术语。
地 理	非洲是世界上磷酸盐和氮的主要出口地区，这是化肥的两种主要成分；除欧洲外，其他各大洲为化肥的净进口地区。
数 学	欧姆定律：$I = U / R$，R是电阻，单位为欧姆，U是电压，单位为伏特，I是电流，单位为安培。
生活方式	人们习惯切菜，可是生菜叶子切开后会释放一种抗坏血酸氧化酶，从而破坏维生素C；我们习惯了有电的环境，电线是我们日常生活的一部分，在我们周围形成电磁场。
社会学	人们认为植物是没有知觉的（当然黏菌也不会有），因为它们没有大脑或神经系统，然而有证据表明植物能够产生反应，还可以学习。
心理学	机器人心理学是研究智能机器性格特征的学科。
系统论	影响土壤电导率变化的因素包括土壤水的连通性以及矿物相的导电性，前者通过土壤密度、土壤结构、水势、沉积物、测量时间、土壤团聚体、土壤水中的电解质等来体现，后者通过矿物质的量、同晶替代的程度以及可交换离子的数量等来影响。

情感智慧
Emotional Intelligence

黏菌

黏菌不确定住哪儿，在找到一个定居之地前向别人征求意见。它对生菜强壮的叶子表示欣赏。它充满好奇心和求知欲，不停发问。在接下来的对话里，它分享了生存和应对食物短缺的经验，为它和生菜如何互相学习提出建议，并期盼有更多对话交流。此外，黏菌还解释了它是有知觉、有记忆的。黏菌提出质疑，如果植物能唱歌，为什么黏菌不可以。黏菌确信它有能力使机器人感到高兴或难过，也能够通过它积极的态度和热情动员生菜与它合作。

生菜

生菜对它的生长环境感到满意，因为那里有充足的养分。生菜有明确的自我认知，它知道自己的能力超出黏菌（以及大多数人）的想象。它不介意黏菌一连串的提问，并且花时间去解释。生菜并不假意想要改变世界，它很谦虚，认为自己的作用微不足道。它对自己以及自己的用处有着详细的了解。生菜把黏菌能够聚合成一个大的个体比作恐怖故事，认为黏菌失控的生长暗含危险。生菜具有怀疑精神。它觉得黏菌有记忆很夸张，后来它意识到自己也没有大脑。尽管如此，生菜仍然不太容易接受黏菌可以唱歌，还可以让机器人感到高兴的说法。生菜最终被黏菌的乐观、热情所感染，相信它们可以一起让世界变得更美好。

艺术
The Arts

黏菌细胞非常智能。你可以为黏菌辟出一块养殖区域，很快这里就会被不同颜色的黏菌所覆盖。你只需提供食物（燕麦粥）。黏菌会在一个平面区域内寻找各种食物，当它发现食物，便会形成一个管状网络结构连接不同食物源。这就是由没有大脑和器官的微生物所创造的生活艺术。你想养殖一些黏菌吗？

思维拓展
Systems: Making the Connections

人类对智能的定义十分明确。我们认为大脑、神经系统和器官的存在象征着更高的发展水平。我们还将痛苦和快乐的感受与大脑联系起来。人类花了好几代人的工夫才认识到动物也是有感知的。这引发了动物福利运动的发展以及停止那些没有必要的、对动物造成疼痛的实验的呼吁。社会正在发掘意识的新界限。虽然大多数科学家认为植物是没有知觉的生物，但越来越多的迹象表明，植物甚至是黏菌都有感知和记忆的能力，还可以创作音乐（见"冈特生态童书"丛书第43册《植物会唱歌吗？》），并使人感到快乐。在过去几十年中，有两万多篇关于黏菌的研究论文发表。实验证明黏菌能做出移动和连接的决定。尽管其行动框架似乎取决于获得食物和繁殖的机会，但有明确迹象表明，黏菌有记忆，可以在迷宫中找到最有效的途径来获得食物。黏菌的本领令我们大多数人感到吃惊，而植物的表现也同样令人印象深刻。植物能够感知到根部周围的障碍物，并在根部碰触到障碍物前就改变生长方向。为防止过度放牧，一些植物会对以它们为食的动物（例如大象）的出现做出反应。大象还没有碰到它们，这些植物就会释放出一种能让大象失去胃口的酸，以此来警告大象离它们远些。尽管这些新的认知尚未得到解释，却可以启发我们用一种不同的方式去看待植物，进而让我们意识到，我们对于智慧生命的认知远远超出我们迄今为止所掌握的。

动手能力
Capacity to Implement

植物色素具有重要的生理机能，如光合作用和抵抗光胁迫。它们的抗氧化性和抗癌作用有益于人体健康。根据一般经验，蔬菜颜色越深，含营养越多。让我们种植不同种类的蔬菜，然后只需要根据它们的颜色，我们就知道哪种是最有营养的。

故事灵感来自
This Fable Is Inspired by

希瑟·巴尼特
Heather Barnett

　　希瑟·巴尼特出生于 1970 年，是一位英国艺术家、研究员和教育工作者，从事艺术、科学与技术的交叉研究。她是伦敦艺术大学中央圣马丁学院艺术与科学硕士课程导师。她经常与科学家、艺术家、其他参与者甚至有机体合作。她的一些作品运用了影像技术与活的素材，包括微生物肖像画、蜂窝壁纸、会表演的墨鱼，还有一个持续合作的伙伴——一种智能的黏菌，名叫多头绒泡菌。

　　她的作品在艺术画廊、科学博物馆和公共场所广泛展出，包括伦敦的维多利亚与艾尔伯特博物馆、科学博物馆、韦尔科姆收藏馆、鹿特丹新研究所以及纽约世贸中心一号楼观景台。

图书在版编目（CIP）数据

冈特生态童书.第四辑:修订版:全36册:汉英对照 /
(比)冈特·鲍利著;(哥伦)凯瑟琳娜·巴赫绘;
何家振等译.—上海:上海远东出版社,2023
书名原文:Gunter's Fables
ISBN 978-7-5476-1931-5

Ⅰ.①冈… Ⅱ.①冈… ②凯… ③何… Ⅲ.①生态环
境–环境保护–儿童读物—汉、英 Ⅳ.①X171.1-49

中国国家版本馆CIP数据核字(2023)第120983号
著作权合同登记号图字09-2023-0612号

策　　划　张　蓉
责任编辑　张君钦
封面设计　魏　来李　廉

冈特生态童书
快乐的机器人
[比]冈特·鲍利　著
[哥伦]凯瑟琳娜·巴赫　绘
颜莹莹　　译

记得要和身边的小朋友分享环保知识哦！
八喜冰淇淋祝你成为环保小使者！